CRYPTOCURRENCY TRADING

How to Make Money by Trading Bitcoin and other Cryptocurrency

by

Devan Hansel

© 2017 Devan Hansel

All rights reserved. No part of this book may be reproduced in any form without permission in writing from the author. Reviewers may quote brief passages in their reviews.

Disclaimer

No part of this publication may be reproduced or transmitted in any form or by any means, mechanical or electronic, including photocopying or recording, or by any information storage and retrieval system, or transmitted by email without permission in writing from the publisher.

While all attempts have been made to verify the information provided in this publication, neither the author nor the publisher assumes any responsibility for errors, omissions, or contrary interpretations of the subject matter herein.

This book is for entertainment purposes only. The views expressed are those of the author alone, and should not be taken as expert instruction or commands. The reader is responsible for his or her own actions.

Adherence to all applicable laws and regulations, including international, federal, state, and local governing professional licensing, business practices, advertising, and all other aspects of doing business in the US, Canada, or any other jurisdiction is the sole responsibility of the purchaser or reader.

Neither the author nor the publisher assumes any responsibility or liability whatsoever on the behalf of the purchaser or reader of these materials.

Any perceived slight of any individual or organization is purely unintentional.

Get the FREE Bonus NOW!

If you're interested in receiving free PDFs on latest strategies, guides and secret tips about topics like cryptocurrency, bitcoin, blockchain, online trading, investing, real estate, stock market etc., I highly recommend you to join my list (link below) I've spent many years understanding all this stuff and I will provide you the distilled knowledge that can not only save you hundreds of hours but also thousands of dollars. Members in my list essentially get to learn how to make money and invest it wisely. To Subscribe, go to the link below.

www.bit.ly/devan-hansel

As a bonus, members will also be getting my latest books for FREE before anyone else. Yes, FREE. However, this is an exclusive invite and will expire soon. It doesn't cost you anything to join. You will only have to put in your email-id so that I can connect with you and keep you updated. It's a clear win-win. So, go ahead and subscribe now by visiting the link above.

Note to the Readers

First off, I'd like to commend you for actually following up on your curiosity and getting this book. Given the public interest and rising market valuation of cryptocurrencies, this book is definitely a smart and timely purchase. The value of cryptocurrencies has skyrocketed since their inception back in 2009 with Bitcoin. A window of opportunity has opened up for those who are interested enough to learn and brave enough to invest. Make no mistake, there is a lot of wealth to be made in this field. 2 years from now, people will look back and wonder why they didn't get on the boat while they still had the chance. The fact that you've bought this book indicates your interest. But are you willing to seize the opportunity?

A lot of time and effort has gone into creating the book you are now reading. And I sincerely hope that it helps you move ahead in your quest for useful knowledge. The book has been designed to take you gradually through the hoops and introduce the cryptocurrency trading landscape, one block at a time. As such, care has been taken to ensure that anybody can read and understand the material without too many prerequisites. The book has also been written in a short-and-concise format so as to allow readers to flip through the book quickly. However, if you happen to find it difficult at times, please go through the resources recommended within the context. Good Luck!

About the Author

Hi there! I am Devan Hansel. I'm a crypto-investor and cat lover. Over the years, I've acquired a wide range of experiences in investing and the art of money-making. I've been involved in the stock market, real estate, startups and more recently…cryptocurrencies and blockchain. Having studied computer science and finance in college, I could easily grasp the essence of the technology and understand how the whole system works. In this book, I've laid out all the essential knowledge you need to understand to start tapping the market and make profitable trades quickly. I've put my maximum effort in making it interesting and understandable. I hope you have a good time reading the book :)

Come join my list if you want to follow latest trends in the marketplace and get huge discounts on early releases of my books. All you need to do is enter your email-id in the link below so that I can communicate with you about my latest works and keep you in the loop.

www.bit.ly/devan-hansel

Table of Contents

Introduction to Cryptocurrency Trading 11

Chatper-1: Overview of Cryptocurrency 19

Chatper-2: The Blockchain Ecosystem 34

Chatper-3: Basics of Crypto-Trading 66

Chapter-4: CryptoTrading vs Forex 95

Chapter-5: Essential tips & Strategies............ 105

Chapter-6: Identifying Profitable Trades 120

Chatper-7: Future of Cryptocurrencies 132

More books from the author 142

Introduction to Cryptocurrency Trading

Nowadays, words like 'cryptocurrency', 'bitcoin', 'blockchain' are being used all around us. They are talked about in board meetings, tech events, major blogs, publications and news channels. But what exactly is all this about? Why are so many billionaires, Silicon Valley icons and financial authorities calling this the next technological and financial revolution? We will learn the answer to this in the upcoming chapters. If you're not familiar with any of the terms or if you're a beginner who is interested in getting a deeper look at this latest phenomenon, you're at the right place...and at the right time too. From the basics of cryptocurrencies to detailed strategies on profiting from the current market trends, you are about to experience the essence of the subject in an entertaining yet educational manner. Brace yourself!

So, what exactly are cryptocurrencies?

To put it short, cryptocurrencies are virtual currencies that use mathematical operations to carry out transactions. They have certain properties like decentralization, durability etc. (we will cover these in later chapters). So, almost everything about cryptocurrencies is digitized including the currency units, regulatory limits, security etc. You can buy a certain amount of cryptocurrency and trade it with others just like you do with the stock market and regular currencies. This is called cryptocurrency trading or cryptotrading for short. The most famous cryptocurrency is Bitcoin which is also the first that was released publicly. Although it's a relatively recent invention, the market scope and investments have grown exponentially. In fact, the current market cap for all tradeable cryptocurrencies is around $621 Billion!

Why should I get into Cryptocurrency Trading?

That's a good question to ask. And frankly speaking, the answer depends on your goals and interests. If you're interested in capturing wealth then the cryptocurrency market is undoubtedly the #1 place to go to right now. Millionaires are being made in a matter of months. This is a fact. You can check out any worthy blog or newspaper or YouTube channel to understand this. And I've certainly made my fortunes with it. Back in 2015, I decided to put in some extra cash from my salary into a few cryptocurrencies. By 2017, my portfolio's net worth was in Six Figures. I cried and laughed as the price went up and up and brought in more than 3000% return on my investment. Was I an exception? Definitely not. There are more than 100,000 Bitcoin millionaires in the world. With basic knowledge of the underlying technology and some market research, anybody can make sensible bets and reap the profits. For additional tips and secret strategies, you are highly recommended to join my exclusive email list. Just go to the link below to sign up NOW.

www.bit.ly/devan-hansel

You might be wondering if market has faded out and if there really is as much opportunity now as there was 2 or 3 years back. If you are, let me make it clear to you. 1 Bitcoin was worth around $19,000 by end of 2017. Experts predict that it will reach a whopping $100,000 (or even more) in only 5 years. This is not a wild guess. It is a well analyzed forecast from the top experts in cryptocurrency. So, 5 years from now, you could be sitting on massive profits or you could look back and regret for not having taken a calculated risk. To help you decide, let me elaborate more. Listed below are the factors that make cryptocurrency trading a sensible option.

1. **Cryptocurrencies are decentralized**. That means, no single financial organization or hedge fund can control the stock value or use market manipulation tricks like *pump and dump* etc. The power is with the public. And the technology that supports this open decentralized framework is called Blockchain which will be covered in chapter-2.

2. **The market is super young.** The first cryptocurrency, Bitcoin, was released in 2009 and trading cryptocurrencies picked up steam around 2012-2013. So, it's only around 5-6 years old. Very few percent of the global population are even aware of cryptocurrencies and even few percent know how to make profits trading them. So, the chances for capturing a piece of the pie are extremely high.

3. **Network effect.** The success of a currency depends on how many people use it and how well it's able to handle transactions. As more people adapt cryptocurrencies, it will speed up the pace of adaption and even more people will learn about them and adapt them. That's the beauty of network effects. If you get paid only in Bitcoins, companies that want you as a customer will be forced to accept your Bitcoins.

Hopefully after reading this, you've got the gist of what cryptotrading is and why exactly it's a gold mine right now. The next steps are to equip you with a shovel and teach you how to dig the gold, metaphorically speaking. Chapters 1 & 2 will teach you the fundamentals of cryptocurrency technology. If you're in a hurry or are already familiar with the details, you can skip chapter-2 which deals with the blockchain technology and why it works the way it does. Chapters 3 & 4 will get you into the shallow waters of cryptotrading and teach you all the relevant terminology. And chapters 5 & 6 will show you the necessary tools and strategies to start making profitable trades. The final chapter will help you understand where the cryptocurrency market is headed in the near future. So, without further ado, let's get started.

Chatper-1: Overview of Cryptocurrency

Before we dive into the nitty-gritty details of cryptocurrency trading, it is important to first understand exactly how they work. Terms like *blockchain*, *mining*, *cryptowallet* will appear frequently in later chapters. You will also need to be equipped with basic knowledge of cryptocurrency ecosystem before you can start day trading or investing in any of the available options. So, let's begin. We will start with the basics and slowly proceed to advanced concepts, tips and strategies.

The Origin of Cryptocurrency

Throughout history, we've used different mediums of exchange like commodity money, paper money, gold standard, fiat currencies etc. But over the years, different scientific communities across the world had been dissatisfied with the short-comings of these traditional currencies. Due to the explosion of internet and progress made in the fields of cryptography, online security, digital payments, it became possible to have a totally decentralized currency that could void the necessity of a central bank or government.

After the 9/11 attacks, America got very strict on the digital front. Laws like the *Patriot Act* were passed to perform online surveillance at a mass level. Needless to say, cryptocurrencies were shunned down owing to their decentralized structure and assumed to be hotbeds for terrorists and other illegal activities.

The first sign of cryptocurrency came when an American cryptographer named David Chaum founded the company *DigiCash* in Netherlands (since it was likely to get shut down in America). DigiCash used *blinding algorithms* to protect user's money and transaction details. However, they had complete monopoly over the supply of the currency and they dealt with the users directly. This made the Central Bank of Netherlands call foul which meant that DigiCash would have to either sell the company or shut it down soon. Although Microsoft approached DigiCash with an offer of $180 million, Chaum thought that it was not enough. So, Microsoft took the offer off the table and DigiCash ran out of funds eventually.

Shortly after that, many cryptocurrency systems like *b-money* and *BitGold* came into light but never took off. They had all the necessary components like blockchain systems, anonymity protection, decentralization etc. but somehow couldn't get enough attention in the marketplace for widespread usage.

The first modern cryptocurrency to emerge that is effective and used widely is Bitcoin. A white-paper explaining the details of bitcoin implementation was first published under the pseudo-name of *Satoshi Nakamoto* in October 2008. The paper is titled *"Bitcoin: A Peer-to-Peer Electronic Cash*

System" and can be downloaded at www.bitcoin.org/bitcoin.pdf. On January 2009, Satoshi released the initial version of the bitcoin software on SourceForge.net, opening the technology up to the public. To this day, the real identity of Satoshi Nakomoto remains a mystery. Based on bitcoin transaction logs, it is estimated that Satoshi owns roughly 1 million bitcoins which are currently evaluated at around 17 billion dollars!

Cryptocurrencies are slowly being accepted by all major companies and startups, especially in Silicon Valley. WordPress became the first major company to accept bitcoins in 2012. Soon after, big shots like Microsoft, Tesla, Dell, Virgin Group, Lamborghini followed. Currently, the total market cap for all cryptocurrencies has exceeded $630 billion. This is an indication that the world is slowly shifting towards decentralized cryptocurrencies for a myriad of reasons.

Different types of cryptocurrencies

More than 1000 public cryptocurrencies exist in the world and many more are created every month. In this section, we will look at the most prominent cryptocurrencies. To view the updated trends and market capitalizations of the top 100 cryptocurrencies, check out CoinMarketCap (www.coinmarketcap.com).

1. **Bitcoin (BTC):** This is the first known cryptocurrency that is well recognized and used by

the public. It has paved the way for modern cryptocurrencies and is considered to be the de facto standard. Almost all the other cryptocurrencies have either branched off from or have major commonalities with bitcoin. Market cap of bitcoin stands at around 270 billion dollars by the end of 2017 making it the largest publicly traded digital currency. For a detailed guide on how Bitcoin works and how to properly invest in it, check out my book *"Bitcoin: The Digital Gold"* on Amazon.

2. **Litecoin (LTC)**: Launched around 2 years after bitcoin, litecoin is a decentralized peer-to-peer cryptocurrency with a growing network of developers, merchants and supporters. Although very similar to bitcoin, it offers relatively faster

transaction confirmations. By the end of 2017, the market cap of litecoin is around 17 billion dollars. Where bitcoin is gold, litecoin is silver.

3. **Ethereum(ETH)**: Launched recently (2015), ethereum is also a decentralized cryptocurrency but offers more functionality like *smart contracts*, the *ethereum virtual machine*, distributed computing etc. As of Dec 2017, Ethereum is the second largest cryptocurrency with a market cap of around 81 billion dollars.

4. **Ripple (XRP)**: Ripple is heavily used by banks to settle global transactions in a secure and effective way at very low costs. It is different from bitcoin in its protocol and structure. Unlike bitcoin, ripple doesn't require high computing power for creation

of new currency. As a result, it has a reduced network latency. The individual units of Ripple currency are called *ripples* (XRP). At the time of this writing, Ripple has a market capitalization of 42 billion dollars making it the fourth largest cryptocurrency.

5. **Dash (DASH)**: Originally known as DarkCoin, Dash is also a decentralized peer-to-peer cryptocurrency like Bitcoin albeit a more secretive one. It was launched in January 2014 and experienced a surge in traffic and fan-following quickly. Its famous features include instant transactions (*InstantSend*) and complete private transactions (*PrivateSend*). It also uses a separate chained hashing algorithm called X11 unlike bitcoin's SHA256.

Note: Cryptocurrencies other than Bitcoin are referred to as "Altcoins" because they are alternatives launched after Bitcoin.

Bitcoin

As you must've already understood by now, the most promising and widely used cryptocurrency is Bitcoin. So, let's look at the factors that make Bitcoin such an awesome currency and why it's a no-brainer to invest in it. For a deep-dive into Bitcoin and how you can potentially make thousands of dollars mining and trading bitcoins, check out my book "*Bitcoin: The Digital Gold*" on Amazon.

Why Bitcoin is a good currency:

1. **Scarcity**: Only 21 million bitcoins can ever exist. We will see why in further chapters. This cap on the total number of bitcoins ensures that its net value never drops too low. As the economy grows, the value of bitcoin also increases. It is estimated that one bitcoin will be worth around 1 million dollars in less than 10 years. And it costs less than $20,000 by the end of 2017. (If you're planning to purchase and invest in bitcoins, check out Chapter-5)

2. **Durability**: The whole purpose of currency is to represent money in a physical/virtual form so that people can have easier time exchanging value. If the currency fades away with time or gets worn

out over repeated use, it can be a hassle to keep churning out more currency to replace the damage. All physical currencies are prone to physical damage like wear & tear, weather etc. This is where bitcoin trumps all other forms of currencies because it is 100% digital. The life-time of a bitcoin is theoretically infinite. It will survive as long as there's an operating network that runs the bitcoin protocol. A decentralized network, high level of encryption, digitized currency and the existential guarantee of internet in the foreseeable future make bitcoin one of the most durable currencies ever created.

3. **Interchangeable**: We already know that currency is just a set of monetary units in use. As such, a good currency is one in which the units are

interchangeable. This means that all the units should be identical in *structure* and represent the same amount of value. Take gold for example. 1 gram of gold has the same value anywhere. Similarly, 1 bitcoin is exactly the same as the other. For all practical purposes, you can exchange 1 bitcoin with another and there would be no difference in value.

4. **Divisibility**: To measure and grade value, a good currency needs to be divisible to the smallest required scale (e.g., dollars & cents, pounds & pence, rupees & paisa). The bitcoin protocol has been designed in such a way that you can divide one bitcoin into many smaller units called Satoshis which can be further divided if necessary.

5. **Transferability**: If you have a working internet connection and a computer (smartphone or tablet will also work), you can transfer bitcoins with just a couple of clicks. This makes it a very convenient mode of money transfer unlike bank cheques and wire transfers. There is no central authority or third party that charges a transfer fee so it is also a more profitable mode of money transfer.

Now that you know what makes Bitcoin a good currency, the next step is to register for a CoinBase account and try buying some Bitcoin. CoinBase is a platform for buying and selling cryptocurrency. It is one of the best options available in the market. I use it myself for all kinds of cryptocurrency related activities and would highly recommend it to you as well. Go to the link below to sign up and get $10 of FREE Bitcoin!

www.coinbase.com/join/598b36cb68284c0125fa0aea

Chatper-2: The Blockchain Ecosystem

In this chapter, we will look at the foundation on which most cryptocurrencies rely on – the blockchain. If you're a beginner to this technology, make sure to take your time while going through the different sections. It is going to get a bit technical. Reference links have been provided where necessary for better understanding. If you're more interested in the trading part, you may choose to skip this chapter and proceed to chapter-3.

What is Blockchain?

Let us begin by asking the question – what do we need a currency for? We need currency so that we can give it to others (buying) or take it from others (selling). Isn't this true? And for a cryptocurrency, that is where a blockchain comes into the picture. Blockchain is a technology that allows people to transfer cryptocurrency between one another securely. It is a distributed database where all the transaction records are saved. Unlike a typical fiat currency like USD, the blockchain of cryptocurrency is distributed and spread across various countries and individuals. The databases and servers are run by volunteers who maintain a peer-to-peer network. There is no possibility of government or any third-party involvement in manipulating the database records. Even if the government officials or any malicious entities volunteer for maintaining the blockchain, they cannot alter the transaction records due to the constraints imposed by its design.

Blockchain is essentially an open electronic ledger where all transactions are recorded for public viewing. These transactions are grouped together into blocks. And as the name suggests, the blockchain is essentially a chain of valid blocks. For example, all the latest bitcoin transactions can be found at: www.blockchain.info. This open strategy of blockchain prevents counterfeits and other frauds. By checking the blockchain, you can be sure that the transactions are completely legitimate. Once you make a transaction, it will appear shortly in the public blockchain.

You might be wondering, *"But won't people know who's spending how much by looking at the blockchain?"*. The answer to that is *No* because your identity is protected using encryption and mapping functions. Only your Wallet-ID will appear in the blockchain which reveals nothing about your personal identity. We will cover the mechanics of a cryptocurrency wallet in the next chapter.

What is Mining?

Now that we have an idea of what the blockchain is, it is necessary to understand how it is updated unanimously throughout the network. It has to be unanimous because if the status of blockchain is not congruent among the nodes, it will lead to discrepancies in verifying transactions which will result in frauds and eventual system failure. So, let's look into this in detail.

There are two kinds of nodes in a blockchain network. Normal nodes and mining nodes. Both of these have their own separate operating protocols. And every node maintains its own blockchain, constructed individually by adding valid blocks to the list. The normal nodes have relatively basic functionality. They receive transaction-messages from neighboring nodes in the network and their job is to verify the transactions and propagate them forward to remaining nodes. This will ensure that as time goes by, only verified transactions are spread across the network. This is a basic layer of security to ensure bogus transactions are not updated in the blockchain.

Now, the mining nodes are a different kind of nodes that execute the *mining protocol* which includes the following steps.

→ Listen for new transactions and verify them.
→ Aggregate verified transactions into a block.
→ Compute the solution to an algorithm called *Proof-of-Work* for that specific block.
→ Timestamp the block along with the computed *proof-of-work solution* and broadcast it across the network.

The mining nodes essentially group new valid transactions into blocks and propagate them to other nodes. The validity of the block can be measured by any node by checking the *proof-of-work* value. So, when this new block is received by the remaining nodes, they check its validity and add it to their own blockchain. A reward is given to the mining node that propagated the valid block with the earliest timestamp. This reward is usually a certain amount of cryptocurrency units. It is similar to the processing fee charged by banks and other financial organizations. Once a block is verified and added to the blockchain, it is said to have been successfully mined.

Now we know how blocks are mined, how the blockchain is built and how the *Proof-of-Work* protocol helps in making sure that every node on the network is on the same page when it comes to the blockchain. Here's the interesting part. In most cryptocurrencies (including Bitcoin), mining is the only way to create new crypto-coins. That is to say, the only way for the system to assign value to the cryptocurrency is to measure the amount of computation performed by the mining nodes. Every bitcoin ever mined has been the result of a mining node performing the *Proof-of-Work* algorithm to create a new valid block.

So, the purpose of mining is twofold. To create new cryptocurrency and update the blockchain with valid transactions. The reason that miners are rewarded is to incentivize them to perform the needed computation. If there was no reward, there wouldn't be enough miners to validate the transactions quickly. This would lead to high latency in the network which would make the whole system rather unsafe. The security of the cryptocurrency depends on how fast the transactions are verified. And that depends on how many miners are competing for the reward simultaneously. This is the beauty of the bitcoin-blockchain system design. And also, the reward for mining goes down by 50% every 4 years for the bitcoin system. Eventually, there would be no reward for mining blocks except for the transaction fee and tips. This is a way to limit the supply of the cryptocurrency and ensure its value doesn't go down below a certain threshold.

The Blockchain is the heart of most cryptocurrencies. It is the bedrock on which all the transactions, security and efficiency of the system rely upon. Moreover, the tech community across the globe is waking up to the ingenuity of the blockchain design. Numerous applications of the blockchain technology are being identified in all areas of the digital spectrum. It may as well be that we've stumbled upon the backbone of a new kind of internet. If you're really interested in learning more about this to get a complete perspective, check out my book *"Blockchain: The Technology Revolution behind Bitcoin and Cryptocurrency"*.

The two main problems that a blockchain solves are:
1. Decentralized Consensus
2. Double Spending

How Decentralized Consensus Works

The architecture of the blockchain is such that it eliminates the need for a central database or monitoring authority. You have to understand that this is a groundbreaking technological revolution not only in the field of digital currency but also in business, banking, governance, politics etc. A plethora of possibilities have opened up after it has been proven that a system like bitcoin can be developed which achieves decentralized consensus in a secure and efficient manner. New self-verifying systems and decentralized apps are being developed today on account of this innovation.

To those who are unaware, decentralized consensus is a scenario where a network of entities comes to a common agreement about something (in our case, the validity of a transaction) without having to trust one another. This is also known as distributed trust-less consensus and is a major research topic in the field of Distributed Systems. Many algorithms have been designed to solve this problem of distributed consensus. Cryptocurrencies like Bitcoin use a specific protocol called *Proof of Work* (POW), as we've seen, which lets the blockchain network achieve distributed consensus and operate without getting tampered with. It is important to understand why achieving distributed consensus is so important in a cryptocurrency's blockchain network.

Let's assume that you have a network of computers (or "nodes") that are interconnected in a haphazard manner. This network forms the *backend* of your service. In other words, all the computation and database storage operations are handled by this network *behind the scenes*. Your objective is to ensure that when a user performs an action, it has to be recorded and updated congruently throughout the network. So, your network is distributed but you have to project a single consistent experience to users everywhere. This is the most basic requirement for not only a cryptocurrency like bitcoin but also for almost every technology company out there like Google, Facebook, Amazon, Instagram etc.

When a user performs an action, you will observe that in order to achieve the objective of consistency, you are inevitably left with only two options. Either record this action in all the nodes or none of the nodes. If you record it in only some of the nodes, there is an inconsistency in the network and the nodes cannot figure out the truth i.e., whether the user did actually perform the action or not. In other words, the network cannot come to an agreeable consensus. This is a big problem because an inconsistent network is an insecure network. Any hacker would be able to exploit this inconsistency to spread viruses or manipulate the database to his/her advantage. Therefore, it is important for a distributed network to maintain consistent data across all nodes and be able to identify erroneous and inconsistent records quickly. This is the reason why distributed consensus is so important in a blockchain network.

Now let's look at how exactly the blockchain aids in achieving this decentralized consensus. We will be considering bitcoin as the reference cryptocurrency.

Decentralized consensus in a blockchain is truly amazing. This is because all the nodes in the network are able to agree on the validity of a transaction without having to trust anyone else or knowing the identity of parties involved. That is why it is also known as trust-less decentralized consensus.

Decentralized consensus in a cryptocurrency using blockchain is achieved in an emergent manner. What this means is that there is no single point of time at which all the nodes in the network are able to agree on the validity of a transaction. As time progresses, more and more nodes will be able to arrive at the same conclusion.

There are four phases in which this emergent distributed consensus is achieved. Let's look at them closely.

Phase #1: Verification of every incoming transaction by every node.

The nodes in the network receive data regarding various transactions from their neighboring nodes. Some of these transactions are just invalid. So, as a primary rule, all the nodes check the incoming transactions and collect the valid ones into what is called as a *transaction pool* or *mempool*. The transactions are verified using cryptographic techniques based on a list of criteria that are public.

And this pool of valid but unconfirmed transactions are propagated across the network by each node. So, all the invalid transactions are weeded out by the nodes in the network in the first phase.

Phase #2: Mining nodes accumulate valid transactions into blocks.

Mining nodes, as we've seen earlier, are special nodes in the network whose job is to collect valid transactions from their neighboring nodes, put them in a block and compute a unique value (*"Proof of Work"*) for the block using a cryptographic algorithm. The mining nodes keep track of the latest blocks and compete with each other to create a new block of these valid transactions with the appropriate *proof-of-work*. These blocks are then propagated across the network to other nodes.

Phase #3: Nodes receive and verify blocks

As the nodes in the network receive blocks from various mining nodes, they calculate their validity. Anybody can accumulate transactions into blocks. But the trick here is that computing the correct *"proof-of-work"* of a block is very hard and therefore reduces the chance of a transaction fraud. Once the nodes receive a mining node's block, they verify it against the *"proof-of-work"* and add it to their blockchain which they've been maintaining and updating so far.

Phase #4: Nodes eliminate irrelevant blocks

Every node maintains and updates its own blockchain which is essentially a list of blocks that are considered valid using publicly known and accepted criteria. Nodes can receive multiple valid blocks by different mining nodes. So how can they decide collectively as to which of the received blocks should be considered while extending the blockchain? This is where the proof-of-work protocol comes in handy. Different blocks have different proof-of-work values. The bitcoin protocol states that while selecting blocks, preference should be given to the block with the highest proof-of-work value. So, if a node gets two different blocks, it will maintain two separate lists in the blockchain until one of them exceeds the other in the total cumulative proof-of-work value sum. It will then discard the sub-chain with lower proof-of-work value sum. In a way, the nodes give preference to the sub-chain in which the mining nodes have spent more computational power because the *proof-of-work* value sum is a measure of the amount of computation done by the mining nodes.

The Double-Spending Problem

The concept of blockchain was first brought to light by the Bitcoin inventor, Satoshi Nakamoto. It was (and still is) considered a brilliant engineering design partly because it was able to solve what no other digital currency could before that point of time which is to *'Ensure that the cryptocurrency units cannot be spent more than once.'* This is termed as the double spending problem.

Unlike fiat currency, the problem with a virtual currency is not the creation of the currency units. Anybody can come up with protocols/algorithms defining how the virtual currency units need to be created, how they need to be structured, what the size (in bytes) of each unit should be and so on. But the fundamental problem that any currency, especially a digital currency, needs to solve is *Double Spending*.

A *double-spend* is a scenario in which one unit of currency is spent in two separate transactions. This can be done by duplicating the unit itself or manipulating the record of transactions. In case of cryptocurrencies, this 'record' is the blockchain ledger.

A typical fiat currency solves the double-spending problem by deploying special techniques to print the cash and identify fake bills. The banks that deal with fiat currency transactions also take extensive security measures to prevent their databases (which hold all the transaction and account details) from getting hacked and hijacked. If the security of the bank's computer-network was compromised, the potential for damage is huge. With countless cases of bank frauds, hacking attacks and duplication of cash, it is evident that a fiat currency's solution to the double-spending problem is undoubtedly flawed.

So, how does a cryptocurrency like bitcoin solve this?

Unlike a centralized fiat currency, a system like bitcoin does not maintain "balances" of the individuals. It only maintains a ledger of transactions a.k.a the blockchain. So, the only way to handle this issue is by assigning identifiers to bitcoins so that when someone tries to spend a bitcoin with the same identifier twice, it can be checked against the transactions recorded in the blockchain.

The way this works is, whenever you send someone bitcoins, that transaction is identified and recorded using a UTXO which is short for Unspent Transaction Output. This UTXO is the unique identifier that represents a transaction of bitcoins which is similar to a bill of fiat currency. UTXOs can be spent only as wholes. But they can be converted into multiple smaller UTXOs for transaction convenience.

When you want to spend some bitcoins, you have to either merge or split two UTXOs to create the new set of UTXOs you want. For example, assume that you have two UTXOs of 0.3 and 0.6 bitcoins, received from Alice and Bob respectively. Let's refer to these using their IDs, X and Y. So, X represents the UTXO of Alice and Y that of Bob.

And let's say that you want to send 0.7 bitcoins to Carter. The conversion goes as follows:

X (0.3 bitcoins) + Y (0.6 bitcoins) => Z (0.7 bitcoins) + W (0.2 bitcoins)

Z and W represent the unique IDs of two new UTXOs created so that 0.7 bitcoins can be sent to Carter. Now, this new UTXO (Z) can only be spent when used in conjunction with Carter's signature. It is propagated across the network and eventually picked up by a mining node which hashes it into a block and updates the blockchain. That is how the transaction takes place. And the conversion is handled by a software called the *cryptocurrency wallet* which we'll be looking into in the next chapter. The other UTXO (W) worth 0.2 bitcoins goes back into your wallet and is spendable only in conjunction with your signature.

With this framework in place, all that a node has to do to verify if a bitcoin is being "double-spent" is to check the UTXO ID against the blockchain's transactions. Even if a node's blockchain is incomplete, the faulty UTXO will get propagated only so far before getting dropped by the other nodes with complete blockchain which can verify latest transactions.

Who maintains the servers and Why?

You must be wondering, if maintaining and updating the blockchain takes so much effort, who would want to do this? Why would anyone want to volunteer for this kind of a task?

The answer to that is *Mining incentives.* As we've already seen, most cryptocurrencies are designed in such a way that the people who validate transactions and update the blockchain are rewarded with new crypto-coins. This serves as an incentive for their efforts. Rewarding the miners is the only sustainable way of maintaining a distributed decentralized cryptocurrency network. This is because mining the crypto-coins requires a lot of computational power provided by specialized GPUs and also involves paying a lot of money in electricity bill.

It also happens to be that mining is the only way of generating cryptocurrency i.e., the new crypto-coins in the network are only generated when a miner creates a new valid block. This is a clever strategy to solve two problems in one shot. The miners get incentivized and the network gets new crypto-coins to work with.

It is very important for the system to be designed in such a way that **anybody** can come in and volunteer as a miner in the network. If the ability to mine was exclusive, the banks or the government or the top 1% could find a way to attain too much control over the system. This could jeopardize the safety and decentralization of the cryptocurrency. For example, if a bank was bombed and/or it's servers were hacked, it's customers would be in trouble. But with a widespread network of mining volunteers, there wouldn't be a single point of failure. This was something that Satoshi Nakomoto made sure of, while designing the system framework.

Why is it safe?

For the purpose of answering this question, let's narrow our focus down to one single cryptocurrency – bitcoin. Bitcoin is the most widely used cryptocurrency in the world. Millions of people pay close attention to the bitcoin network every day. The software itself undergoes regular public updates. A 2013 article on Forbes suggests that the global bitcoin computing power is 256 times more than the top 500 supercomputers in the world. That should give you a measure of the number of servers being run by the bitcoin volunteers. So, at this point of time, the only possible ways to hack bitcoin are either taking down the internet or cracking the SHA256 function. SHA256 is one of the most famous security functions used in the Bitcoin protocol to encrypt data into output of 256-bits (32 bytes) size. It is deemed to be uncrackable. This of course, is regarding the overall bitcoin network and the system design. You can still get your bitcoins lost/stolen if you do not follow the recommended security measures (described in the next chapter) while operating your wallet.

Although cracking the SHA256 algorithm is next to impossible, it is important to understand the difference between safety and anonymity in the context of cryptocurrencies. The fact that SHA256 is hard to crack only implies that an attack on the blockchain or stealing your crypto-coins is extremely unlikely. It, however, does not mean that you are anonymous within the system. This is one of the biggest misconceptions about bitcoin and other cryptocurrencies. Your bitcoins are safe but your identity is not a total secret.

Most cryptocurrencies including Bitcoin only provide pseudonymity and not complete anonymity. Although your identity is not revealed openly, the transaction details are updated on the blockchain which is accessible by anyone. Using techniques like cluster analysis and pattern recognition on the data from the public blockchain, one can start to form associations with your activity and your IP-address (which is essentially your online identity). Now, you can use software like Tor or VPN to hide your IP address but the fact of the matter is that even Tor cannot guarantee complete anonymity. A dedicated hacker with enough resources can eventually track your IP address down. But, he/she will not be able to steal or tamper with your crypto-coins as long as you follow the proper security measures and store your crypto-coins in a safe wallet. Having said that, if you're still concerned, I would recommend that you choose Zcash as it is the most pseudonymous cryptocurrency out there.

Chatper-3: Basics of Crypto-Trading

There is a lot of euphoria in the crypto market currently. And that calls for caution. As Warren Buffet said, *"Be fearful when others are greedy and greedy when others are fearful."* And the recent upsurge in the valuation of cryptocurrencies indicates a lot of greed in the market currently. People are buying in like crazy and many are holding on to what they've got. There is a lot of hype going around. This is the time for caution and paying careful attention to the market while getting educated on the subject. And that's what we're going to do. In the next few sections, we will look at all the basics of crypto-trading and understand terminology like crypto-wallet, crypto-exchange etc. We will also be looking at some of the top cryptocurrencies available in the market and what they have to offer.

How to store cryptocurrencies

For regular fiat currencies, we all know where to store them i.e., in banks and wallets (online/offline). But how do you store cryptocurrencies? And how do you ensure that they're safe from thefts and attacks? We will look at how these problems are handled by a software called the "wallet" in the next section.

Cryptocurrency wallet

A cryptocurrency wallet, or *crypto-wallet* for short, is a digital holder for your cryptocurrency (like bitcoin) and is mandatory for performing transactions. It stores your private, public keys and manages your cryptocurrency transactions by interacting with the blockchain. There is no such thing as a bitcoin without a wallet identification. Every cryptocurrency unit has to be associated with, and transacted using, a wallet. You cannot spend your crypto-coins without the wallet. You also cannot spend the same crypto-coins from multiple wallets because it doesn't tally with the blockchain's record.

There are different types of wallets you can use. You will find below, an image of a mobile wallet which is essentially a mobile app that stores your public & private key data and manages your transactions. The specific screenshot has been taken from the *Bitcoin Wallet* app on Google Play Store. There are other types of wallet frameworks like desktop wallet application, online wallet(website), hardware wallet (USB drive, hard disk etc.), paper wallet (printed sheet of keys in a QR code).

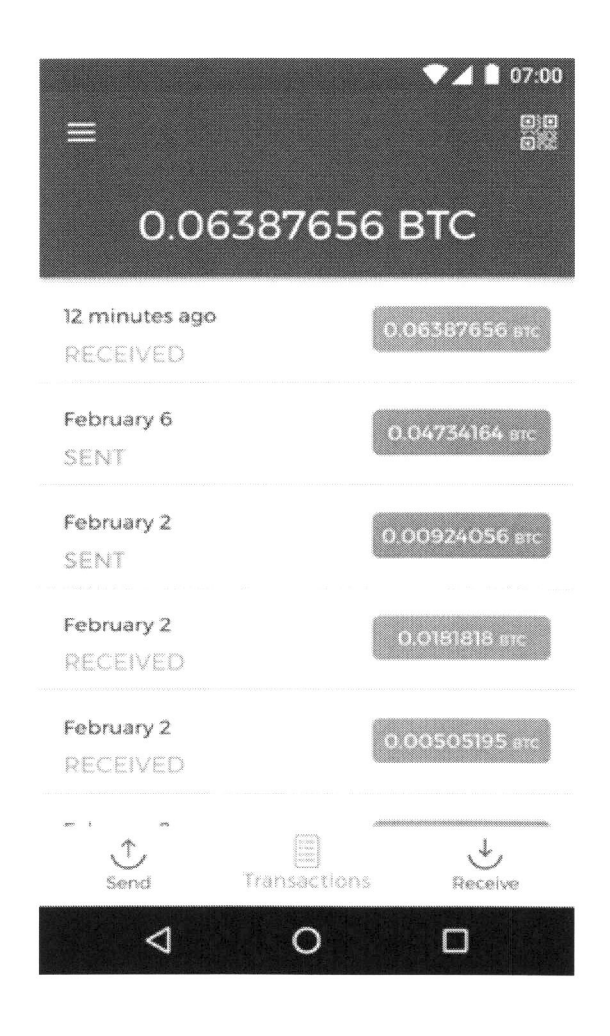

How does the crypto-wallet work?

A cryptowallet holds 3 primary values. The public key, private key and the amount of crypto-coins.

As we've seen, the primary purpose of a wallet is to facilitate cryptocurrency transactions. Here's how it does it.

Let's say that you want to send bitcoins to your friend. Your wallet will generate the transaction-message, number of bitcoins you want to send and sign it with your private key and your friend's public key. This message is then communicated over an online network channel with your friend on it. Your friend's wallet will verify if the message is in fact sent by you and intended for him by decrypting it with his private key and your public key.

After the authenticity is established and the possibility of a middle-man is eliminated, your friend's wallet increases the number of bitcoins it holds and sends a response. Once your wallet receives the response, it decreases your bitcoin amount.

There is a specific Wallet protocol put in place to ensure that the amounts in the two wallets corresponding to a transaction are modified correctly. Anybody who wishes to implement their own wallet software must adhere to this protocol or else the transactions won't be processed.

After the amounts in both the wallets are modified, the blockchain is updated with the transaction entry. It takes some time for the blockchain network to validate the transaction. If all goes well, the ledger moves forward otherwise the error in the system will notify the wallets and the change is reverted. This concludes a typical wallet use-case scenario.

For the purpose of simplicity, many details have been omitted. If you're looking for more specifics, please visit the official bitcoin developer guide on this topic.

Security Measures for Wallets

Losing the wallet or the keys will result in ***TOTAL LOSS*** of your cryptocurrency. It might helpful to learn about a famous real-life story of James Howells who lost 7500 bitcoins (worth around $120 million today) because he accidentally threw his old hard drive into the trash bin while clearing his desk. That hard drive is now reportedly buried under four feet of junk in a landfill site in Newport. So, make no mistake, the security of your wallet should be your top most priority when dealing with cryptocurrency. Here are some tips to follow.

Tip #1: There are different wallet software you can choose for any cryptocurrency. Please use only an officially recognized wallet to avoid issues of security and malfunction. Take some time and go through the wallet specifications and your cryptocurrency's website to pick what's best. For bitcoin, you can find all the recommended bitcoin wallets at: www.bitcoin.org/en/choose-your-wallet.

Tip #2: Encrypt your wallet and private key and have multiple copies stored in secure locations (online and offline). Make sure that you have at least one copy available in an accessible physical device like a flash drive or a hard disk.

Tip #3: If the amount of your cryptocurrency is substantial, it is recommended to use multiple wallets to distribute the coins and reduce the possible damage that can happen. Use 2-step verification methods or MultiSig (Multiple Signature) transactions.

Cryptocurrency Exchanges

Also called *crypto-exchanges*, these are online platforms for buying and selling cryptocurrencies.

You need to connect your wallet with a crypto-exchange to start trading. You can also "buy-in" with your fiat currency after verifying your identity i.e. you can purchase BTC for, let's say, USD.

Just like regular company stocks can be traded at stock exchanges like NASDAQ, NYSE etc., cryptocurrencies can be traded at these crypto-exchanges. For a list of the top crypto exchanges, check out CryptoCoinCharts at the link below. www.cryptocoincharts.info/markets/info

Please note that it's not necessary for an exchange to support all cryptocurrencies. And some of them might not be supported in your geographical area. So, browse through the exchanges carefully and select one that you find suitable. Here are some parameters to judge the exchange on: reputation and public opinion, supported payment options, transaction fee, geographical limitations, supported cryptocurrencies, ease of usage etc. (we will cover these shortly). The most popular exchanges are CoinBase and Kraken.

I personally use CoinBase (link below) because it satisfies all the essential criteria and offers a top-notch customer service. And so far, it has been a safe and smooth ride without any issue. I highly recommend it to beginners and anybody interested in cryptocurrency trading who wants a pleasant trading experience. Use the link below to sign up and get $10 bonus for your first trade.

http://www.coinbase.com/join/598b36cb68284c0125fa0aea

After picking a suitable exchange, you will need to verify your identity (via passport, driver's license etc.) to create an account. Once the account is created, you will be able to add/withdraw funds and start trading. Just like any other trading platform, you will be charged a very small fee for every trade to keep the exchange going.

You might be wondering as to why your ID is required when after all, cryptocurrencies are supposed to be decentralized and support users' privacy/anonymity preferences. Well, here's the thing. Although the transactions themselves are private, the cryptocurrency exchange needs initial fiat currency funds to assign you crypto-coins to trade with. And where there is fiat currency involved, there is a non-zero probability of financial fraud. So, to avoid issues with unoriginal fiat currency (stolen credit cards etc.), the exchange does require your personal information to validate your fiat money. Once you've been verified, you can trade on the platform with privacy.

There are a few types of crypto exchanges that exist out there. So, it might help you to be aware of them before getting your feet wet.

1. **Traditional Crypto-Exchanges**: Similar to the old-school stock exchanges, these act as the "middle man" for traders looking to buy/sell

cryptocurrencies at market price. A slight fee is charged for every transaction to keep the exchange going. They also let users "buy in" with regular fiat currencies. Popular examples are: GDAX, Kraken, Shapeshift.

2. **Direct Trading Exchanges**: These are a kind of "unofficial" platforms where the trade doesn't happen at the fixed market price. Instead, sellers can set their own price and trade directly with buyers. Also referred to sometimes as peer-to-peer exchanges.

3. **Crypto Brokers**: These are independent platforms that offer cryptocurrency trading, customer support, development and other services. Similar to the currency exchange booths

at airports. Designed for a smooth trading experience, they provide trust and support to intermediate and advanced traders who find the traditional exchanges lacking in proper user interface and/or functionality. A good example for this is Coinbase.

Now let's look at the factors to consider before selecting a cryptocurrency exchange to trade with. This is an important part of the trading process since you will be holding your crypto-coins on the exchange, sometimes for long periods of time. It's better to be safe than sorry. With that, let's dig into what exactly makes a good crypto-exchange and how to check if it suits your trading needs and preferences.

- **Credibility**: Do your research before selecting an exchange. When you find something that looks good, ask around and investigate a little. There are a lot of good forums that can assist you like BitcoinTalk (www.bitcointalk.org), CryptoCompare (www.cryptocompare.com), Reddit (www.reddit.com/r/CryptoMarkets) etc. You can also check out the latest Google News articles, Quora answers and do a general web search to get an idea of the exchange's trustworthiness.
- **Payment method**: Does the cryptoexchange accept your preferred method of payment? Are other options like PayPal, credit/debit cards, wire transfer also accepted in case you want to change your mode of payment? On

a side note, cryptoexchanges usually charge an additional premium for credit cards due to the added risk and processing fee.

- **Identity Verification**: This is a sort of protection mechanism for the exchanges to prevent users from trading with black money and other frauds. But always be careful when submitting your personal information to the exchange. Uploading a government-issued public ID like Driver's license or Passport should be fine.
- **Fees Structure**: This is something that changes with every cryptoexchange. You are usually charged for either deposit, withdrawal or transactions. Exchange rate is also something that influences your trades especially when done in large amounts. The

exact percentages vary widely so you should always check the cryptoexchange's website for complete details.

- **Geographical Restrictions**: This is again one of the factors that changes a lot between exchanges. You should check if the exchange offers *full* set of features and services in your country. Also, if you plan to do trading while on vacation in another country, you should check that as well.

Now that we are well versed in the methodology behind trading like how cryptocurrencies are stored, how exchanges work, what to look at while picking an exchange etc., let's understand at a fundamental level, the answer to the following question.

Where do cryptocurrencies get value from?

One of the original reasons cryptocurrencies were invented is to store digital assets securely and avoid interference from central powers like the governments and banks. Some cryptocurrencies are backed by gold and precious metals while others have no backing except the widespread acceptance by users. So, if there is no backing from the governments and everything is distributed globally, where do cryptocurrencies actually get value from and what are the factors influencing it?

1. **Supply and Demand**: One of most popular economic principles is the correlation of price of an object with its supply & demand. Let's take

Bitcoin for example here. As we've already covered, there can only be 21 million bitcoins in circulation due to the mining constraints. There are 7 billion people on this planet and as the adoption of Bitcoin as a global currency grows, there will be friction in the market caused due to growing demand and increasingly limited supply. This friction will lead to the rise in value of Bitcoin. It will also be amplified due to the fact that the popular strategy among many crypto-investors seems to be to "buy and hold". We will look at trading strategies in a later chapter.

2. **Mining difficulty**: Unlike fiat currencies which are minted by the national governments based on various monetary policies, cryptocurrencies are mined by volunteers. Mining cryptocurrencies

requires a lot of electrical and processing power. And the cost is not getting any cheaper. So, the inherent difficulty involved in creating a unit of cryptocurrency leads to a certain perceived value. This goes up as the mining difficulty increases. Basic economics states that the price of something that is rare and valuable will be high. Many cryptocurrencies use POW (Proof of Work) protocol while mining new coins and validating transactions. And the POW protocol rewards miners who have spent more time and effort solving harder problems. This is a fair way to incentivize the miners and also ensure that the price of the cryptocurrency and mining difficulty are directly related. For more details on this, check out my book *"Blockchain: The Technology*

Revolution behind Bitcoin and Cryptocurrency" on Amazon.

3. **User Requirements**: If a cryptocurrency has no practical benefits to users, why is it any good? The cryptocurrency has to solve user's problems to be considered valuable. Similar to how a company's stock value will plummet if it's not delivering any good quality products/services to its customers. In addition to being a means of exchange of value, many cryptocurrencies offer distinctive solutions to domains like legal contracts, digital security, Internet of Things etc. This makes the investors fund the project and the customers register and use it.

4. **Public Opinion**: This is one of the most underrated causes of price surges for not only cryptocurrencies but any other publicly traded stock/commodity. Despite what we may believe, a majority of people think emotionally and take decisions based on their gut. The term "panic selling" is famous among traders. Any major news like a security breach or a market crisis will make the value drop. This happened in Feb 2014 when Mt.Gox, the most famous crypto-exchange at that time filed for bankruptcy as a result of cyber-attacks. When it comes to Bitcoin, a lot of people believe that the independent decentralized nature of currency will be more beneficial since it is less prone to corruption, fraud and manipulation by central banks and governments. The other side of the coin (no pun intended) is that there are also a

lot of people who believe that Bitcoin is a currency used mostly by drug-dealers and criminals online. Nevertheless, the fact of the matter is that cryptocurrencies are blowing up globally and more people are becoming aware of the current crypto landscape and what it has to offer.

5. **Media & Law:** The price of a cryptocurrency can rise or fall depending on how the media portrays it to the public. There is always the possibility of getting blindsided by manipulative media. A few corporations or individuals who hold vested interest in a cryptocurrency can publicize its ICO (Initial Coin Offering) to bloat up its price in the market. This is why you should always do proper market research and look into a wide variety of sources including reddit forums, quora answers,

facebook groups, google news/trends and multiple news and publishing articles. If you're tech-savvy, I would also advise you to delve into the source code and developer updates. Legal notices, Nation-wide bans and anti-cryptocurrency policies have also been observed in the recent past. Countries like China, Vietnam and Russia are active in their protest against public usage of bitcoin. This caused a temporary dip in the price of bitcoin but soon bounced back up to an all-time high. Meanwhile, many countries like Canada, UK, Australia are embracing the crypto-revolution and have provided infrastructure and policy support to the cryptocurrency communities. Some of them even have Bitcoin ATMs available across various cities. You can check out the article below to get more details on this.

https://news.bitcoin.com/worlds-top-10-bitcoin-friendly-countries/

6. **Investors**: The fact that a cryptocurrency startup has received funding from a good investor can boost its coin price in the market. Investments are generally considered signs of trust. So, when a good/popular investor decides to put in capital for growing a cryptocurrency, a large portion of people also decide to place their bets on it as well. Some malicious investors can also try to buy a large portion of the coins, inflate the price with press-releases or promotions and then sell them off quickly without any real progress. This is also referred to as the *pump and dump* strategy. The investors thus have a considerably high impact on

the pricing of cryptocurrencies (especially altcoins with lower market caps) indeed.

7. **Market dilution**: With more than 1000 cryptocurrencies currently in the market and more coming in every year, the market sure has gotten crowded with so many alternatives. Even if an innovative solution is offered by a brand-new cryptocurrency in the market, it doesn't take too long before a competitor opens shop with lesser token price and upgraded capabilities. This causes frequent and unexpected spikes in the prices of cryptocurrencies. Bitcoin, though, is considered a reserve cryptocurrency since it has the highest market cap and largest user-base, owing to its first mover advantage. Fluctuation in bitcoin price usually causes a ripple effect and creates a

fluctuation in prices of other cryptocurrencies as well.

Chapter-4: CryptoTrading vs Forex

Let us begin this chapter by studying the basic trading terminology that is used not only in cryptotrading but also in other platforms for stock trading, forex etc.

Trading Terminology

The basic theme of trading is buying and selling. And when it comes to that, there are a lot of price points and values that determine where a certain cryptocurrency is bought and sold at. For beginners, here are the most commonly used terms when dealing with various prices in market.

ASK: This is the price that the seller asks for the cryptocurrency to be sold at. Anything lower will not be accepted by the seller. This ASK price will vary based on the demand for the cryptocurrency in the market i.e., ASK will decrease if nobody is interested in buying it. So, the price at which a certain cryptocurrency unit is sold will always be higher than or equal to the ASK price.

BID: The BID value is the highest price that a buyer is willing to pay for the cryptocurrency. If the ASK is the lower limit on the selling price, think of the BID as the upper limit on the buying price. Anything higher will not be accepted by the buyer. The BID price will rise as the demand for the cryptocurrency increases in the market. The default BID is the price at which last successful sale was made.

OFFER: Cryptocurrency can be bought and sold at various online platforms or even directly in a peer-to-peer fashion. So, we need a standard buying and selling price to get an idea of how far off from the average we are at. For this, we use the cryptocurrency exchanges where trade volumes are very high. The OFFER is the price at which the exchange is willing to sell the cryptocurrency. It is usually the highest BID price and does not include the commission/fee that the exchange charges for every trade.

BUY: This the price at which the cryptocurrency is actually sold at. The trade happens at the BUY price. It can also be seen as the BID price that matched the OFFER price.

These four price points i.e., ASK, BID, OFFER, BUY will be the guiding benchmarks throughout your trading journey. You need to study them for various trades to understand where the market is at and how the demand and supply change.

Hey there! Just a quick break before we continue learning further. How do you feel? Are you already familiar with the concepts described here? Or are you enjoying learning about these new ideas and methods? Please let me know. I'd like you to write a review on Amazon by using the link below. It will take no more than 2 minutes and would mean a lot to me. Thank you ☺

www.bookstuff.in/devan-hansel-cryptotrading-review

Now let's look at how different cryptocurrency trading is from forex trading. Both of them have their own pros and cons so we will look at each of them in comparison.

Pro Cryptocurrency

1. **Very easy to get going.** Unlike traditional Forex trading which may take few weeks to get started with, cryptotrading literally takes less than an hour to get started. For forex trading, there are a lot of hurdles on the way like signing forms, getting access codes from the broker etc. before you get to transfer your money from bank account and start trading on the exchange. But with

cryptotrading, all you need to do is sign up with an email, upload an identity proof and you're done!

2. **Very easy to quit.** Unlike forex trading which is usually a big pain to exit (broker issues, tax etc.), cryptotrading is super easy to stop when you want it to. You just have to transfer your cryptocurrency from the exchange onto your wallet. Easy peasy lemon squeezy.

3. **Smaller Spreads.** The difference between ASK and BID price is called spread. This is considerably small for cryptocurrency trading when compared with forex. That means, when you exchange a cryptocurrency like Bitcoin(BTC) with another cryptocurrency like Ethereum(ETH) or even

regular fiat currency like American Dollar(USD), you have very minimal loss in terms of value. But it's not so with forex. For example, you exchange USD with EUR and exchange that with INR and repeat this exchange of currency multiple times. Suppose you end up with USD as the final currency. Because of the difference between the ASK and BID prices during these currency exchanges, you will end up with a USD amount that is lesser than what you started with. And this difference (caused by higher spreads) is not insignificant for forex while it is very small for cryptocurrency trading.

4. **Inflation.** Many crypto-enthusiasts and experts believe that Bitcoin is immune to inflation because of its limited supply. Only

21 million bitcoins can ever exist because of the mining algorithm. Since you can't mine bitcoin after this limit, it will be immune to debasement or monetary inflation. This is not the case with fiat currencies because the government can issue orders to mint money depending on the nation's economic circumstances. So, while trading cryptocurrencies, you have one less factor to worry about i.e., you don't need to account for monetary inflation (although price-value inflation is still possible but that's not too harmful).

Pro Forex

1. **Volatility.** The volatility or rapid-change factor for Bitcoin is 10% average while it is only 0.1% even for the most differentiated currency couples in Forex. To put it another way, forex trading doesn't involve overtly rapid change in the currency values while cryptocurrency trading does. As a result, cryptocurrency trading attracts investors who like higher risks (as compared to forex). However, if you follow some of the strategies (like buy-and-hold) discussed in the next chapter, you will be at no significant risk due to the volatility of cryptocurrencies.
2. **Predictability.** Unlike cryptocurrencies whose value is just based on the demand and

supply, forex currencies are more predictable by knowledge-based analysis. The correlation between various geopolitical and economic factors with the currency value is more direct in forex.

3. **Stability.** A cryptocurrency may disappear over time if nobody is interested in it. However, this doesn't happen (so easily) with regular fiat currencies because the economy of a nation will depend on the performance of its currency. So, as a forex trader, you can be sure to have a relatively more sustainable journey when doing long-term forex trading as opposed to long-term cryptocurrency trading.

Chapter-5: Essential tips & Strategies

Cryptocurrencies especially Bitcoin have had mass adoption unlike any other in the fintech space. Celebrities and top Investors like Mark Cuban, Peter Thiel, Bill Gates, Dan Bilzerian, Richard Branson have publicly proclaimed their interest and hopes for Bitcoin. It is a fact that cryptocurrencies have had some of the best ROIs (Return of Investment) ever in history. Many people have made more than 3000% profit in the crypto market. But it's also true that there are many cases of people losing all their savings on cryptocurrencies like Ethereum or Dogecoin. As for me, I've personally made over six figures in cryptocurrency trading and have plans to invest long-term in this field. The reason for this is that I believe the cryptocurrency technologies (including blockchain) will have an unprecedented impact on the way we deal with money and handle digital transactions. Apart from the techniques we will learn in the upcoming chapters, this mindset of long-term investment and stable approach (as opposed to the get rich quick pump-and-dump strategies) has helped me make my profits. And that would be my advice to you as well. Only invest with money you can afford to lose and develop a calm stable long-term attitude.

There are two main strategies when it comes to making money with cryptocurrencies.

1. Buy and Hold
2. Day Trading

The *Buy and Hold* strategy is what seems to be the most popular among beginner crypto-traders. As the name suggests, it involves buying a high-traffic cryptocurrency like Bitcoin or Ethereum and holding it long-term (1-5 years) in order to reap the benefits later. It's like a 'set it and forget it' type of situation. Once you hold the stock, you don't even check the news or get any updates regarding the price surges. This is because the *buy and hold* strategy requires you to be impervious to impulsive decision-making. On the off chance that the cryptocurrency you're holding is blowing up in price like crazy and you want to capture some of that, you may decide to sell some of your stock in installments. But make sure to hold the majority of it in your wallet for the long-term.

On the other hand, *Day Trading* is for those who are actively involved in the market, looking to profit by buying low and selling high in relatively shorter time periods. Unlike what you might have read or heard, day trading is tough and frustrating. You have to study a lot and struggle to keep a clear perspective. Day traders are like whale hunters, trying to bet against the majority and tackle the market for profit. So, I just want to re-iterate that day trading may not be for everyone and that's fine because there are many other ways of making money with cryptocurrencies like *buy and hold*, investing in ICOs, mining etc. (we will cover these shortly). However, if you do decide to take this route, here are some tips to guide you along the way.

Tip #1: Protect yourself from FOMO

Short for Fear of Missing Out, FOMO is one of the biggest enemies for people in all kinds of areas like trading, investing, entrepreneurship etc. Day trading is usually done in certain time periods where the markets are open and ready. If you miss the window for a day, don't sweat it. You can ride the wave tomorrow. However, if you get anxious and try to recoup your share of the possible profits, you will most likely end up losing what you have. Getting caught up in the hype is another thing you want to avoid. For example, back in 2013, there was a kind of frenzy going on with the bitcoin market. Everybody started hyping up the possibilities of bitcoin and how it can solve all kinds of issues like world hunger and poverty. In addition to that, because of China's new policy on Bitcoin, the price got bloated up to around $970. I knew that many people would've fallen victim to FOMO so I held my money and observed in silence. Soon enough, the mini-bubble popped and the BTC price came crashing down to around $430. That is where I entered the market and bought in while others were selling off in haste to avoid further losses. The price of 1 bitcoin has since moved up from approx. $430 to $19,000 (Dec 2017). So, it goes without saying, protecting myself from FOMO has made me more than Six figures in profit while it has left countless others in regret and self-pity. Day-Trading with cryptocurrencies requires you to think rationally and get rid of any false beliefs and/or fears. This is the single most valuable advice I can ever give you.

Tip #2: Plan before you trade

This may seem like a pretty basic advice but I'm astounded how many people forget to implement this properly. The majority of traders forget to plan so they throw money into the market and *hope* to make some profits. This is not even an exaggeration. The data has repeatedly shown that peaks in the bull phase (rising value) are usually higher than expected because people get lured in by the hype and forget to set a proper entry and exit points. Over 90% of people buy when the stock price is high because nobody wants to buy it when the price is low. They need the stock to get hot before they can believe that it's a good trade or investment. So always prepare a well-researched plan and stick to it.

Tip #3: Do your own research

There are many traders and platforms online that try to do your homework for you. They will set the target selling/buying price for a cryptocurrency and predict that it will be profitable based on the trends and other analytics. One of the most famous among such traders is the *Wolf of Poloniex* (https://twitter.com/WolfOfPoloniex). I'm not saying that you shouldn't follow them or heed their counsel. What I'm saying is that you should *always* do your own research before making a bet. Don't let anyone ever convince you that you should follow them blindly. Given the volatility of markets at times and the impulse to get rich quick, it is extremely easy to fall into the trap of letting other authorities dictate your trades. In the long run, developing self-reliance will help you take better decisions and deal with their consequences.

Tip #4: Stop over-trading

This is more prominent among beginners and over-enthusiastic (borderline desperate) traders trying to squeeze the market for all the profits it can provide. The reality is that you only need one single trade/investment at the right circumstances to make all the profits for the year. The problem with over-trading is that your profits are marginalized due to the exchange fees that gets added up over the numerous trades. To minimize this exchange fee and commissions that trading platforms charge, you need to reduce your trade-count to an optimal minimum.

Tip #5: No Regrets

Whether you're buying or selling cryptocurrency, you should always ensure that you're OK with the consequences. No matter how the price fluctuates, you should come out of the trade with the maximum possible contentment. How do you practice this? Well, the first thing is to only invest with money you're OK with losing. This is one way to avoid regrets when the stock tanks and you lose the money (unless it's a long-term strategy). You cannot force the market i.e., the factors that define whether the price will go up or down are beyond your control. The earlier you accept this, the smoother your trading process will be and the fewer regrets you will have. Another way to avoid regret is to sell a portion of stock when the price goes up really high. This is because you cannot predict if the price plunges down in the near future. So whenever possible, you want to capture the profits for a rising stock. Of course, it goes without saying that you should only do this when the profit margin is really high like 300% or more. That way, even if the market crashes, you'd have made your margins before that happened.

Cryptocurrency CFDs

Contract for Difference(CFD) is a contract between the buyer and seller in which the seller agrees to pay the value-difference of the product from the time of selling to a certain other time in the future. You can make money dealing these contracts at various CFD platforms available online. They are highly regulated by the government and have a relatively stronger guarantee when it comes to safety & security of your cryptocurrency. You also won't be needing any additional crypto-wallet to make these contracts. However, the fees can be much higher than regular cryptoexchanges.

Cryptocurrency Leverage Trading

Leverage trading is basically a traditional stock market trading strategy where you can borrow some funds to increase the capital that you can trade with, thereby giving you a "leverage". It is expressed as a ratio. For example, if you have an amount of $1000 in your account and get a 5:1 leverage, you will now be able to access $5000 for trading. The loan is given to you by a broker and is called "margin". You will get this margin when you have enough cash/securities in your account to suffer any collateral damage. And the fundamental reason for leverage trading is to make larger profits on smaller price-changes. One thing to keep in mind is that leverage trading (also referred to as margin trading) has high risk involved with it. The leverage you get can incur you higher loss than you would've got without it.

If you are someone who has limited cryptoassets and would like to acquire more resources, you can opt in for margin trading with any of the cryptoexchanges that support it and start trading with leverage. And if you are someone who wants to lend out extra cryptocurrency, you can do that as well. While you provide leverage for other traders, you will be able to benefit on the loan's interest which is usually provided by the exchange itself. For example, with Poloniex, you can lend bitcoins or altcoins and get a steady interest on your loan provided you hold the coins in poloniex's wallet. You should be careful and read the appropriate guidelines and terms before opting in for margins in a cryptoexchange.

Liquidation value: While trading on leverage, if your account reaches a value so low that you won't be able to pay back the loan, the cryptoexchange will take control and halt you from creating further losses. This is called the liquidation value. You can keep trading after this point but any profits/losses will be made from your own coins.

There are many advantages to cryptocurrency leverage trading. You can hold minimal cryptocurrency on the exchange and make great profits on borrowed coins. Cryptoexchanges like BitMEX offer up to 100:1 leverages. BitMEX is one of the most reputed cryptoexchanges that is favorable for leverage trading. It has a great support team and is super easy to use.

Chapter-6: Identifying Profitable Trades

One of the most important aspects of day trading cryptocurrencies is market research. You need to be able to analyze the data and explore the buy/sell trends. The first thing you need to do is look at the price graphs and get a sense of how well the cryptocurrencies are doing. You can view the performance graphs of the top 100 cryptocurrencies at CoinMarketCap (https://www.coinmarketcap.com/)

Go through the list and filter out at least 5 cryptocurrencies that are showing a decline in price in the recent past (approx. 10 days). You can use additional criteria like total market cap and total traffic volume (from google trends) in your filtering process.

Once this step is finished and you have identified potential candidates, you want to analyze if there are any "buy walls" or "sell walls" i.e., if there are any obstacles in buying or selling the cryptocurrency at a mass level in the near future. You can do this, for example, by calculating how much bitcoin it would take to double the value of the cryptocurrency given its current total trading volume. The key idea here is to look for altcoins that do not have significantly high sell walls so that the probability of the value rising in the near future is reasonably high.

Although Coinbase and CoinMarketCap are good places to start from, if you're interested in digging deeper and want to pick up on more subtle market trends, GDAX is the place to go to. GDAX (Global Digital Asset Exchange) is a trading platform for advanced Coinbase users. They are mutually compatible so you can use the same login credentials on GDAX as Coinbase and also transfer cryptoassets between them. This is how the GDAX home page looks like.

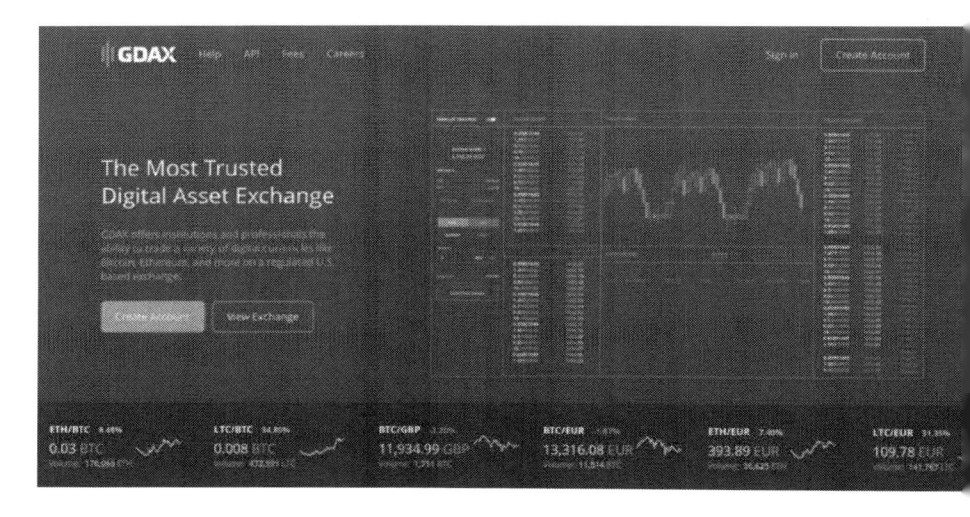

Figure 1. GDAX Home Page

Upgrading to GDAX will give you a lot of benefits including a lower transaction fee on trades and other additional features like real-time pricing data, simple visualizations and charts, order book, trade history etc. But the only downside is that there is no guarantee for crypto-sellers that their order is going to get fulfilled. So, there is a probability that your price for your cryptocurrency may not be met by the market. In that case, you will have to lower the price.

Here are some major features of GDAX that will help you identify trends in the cryptomarket.

1. Price Chart: This is the most famous and most used chart of GDAX. It shows the pricing and volume data of the cryptocurrency over time. The candlesticks in the chart show the open point and close point of the particular time period.

2. Depth Chart: Present just below the Price Chart, the Depth Chart (figure 2. below) will help you identify the demand and supply of the cryptocurrency. The chart consists of essentially two graphs. One for the BIDs (Buy orders) and one for the ASKs (Sell orders). We have already discussed about BID, ASK etc. in the previous chapters. In the depth chart, the green portion (left side) indicates the buy orders a.k.a demand and the red portion (right side) indicates the sell orders a.k.a supply. For example, you can select a point on the ASK graph and know how many units you can sell (cumulatively) at that particular price point.

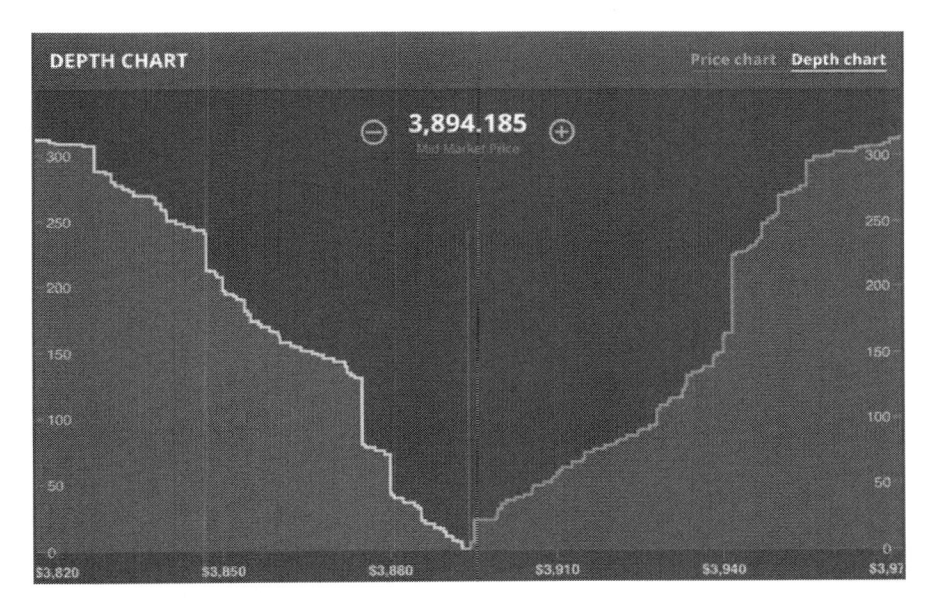

Figure 2. Depth Chart

3. Order Book: This is present on the left side of the price chart. It's like a live counter, showing all the current open orders on the GDAX+Coinbase platform. You can click on an order and choose to buy/sell the same amount for your own order. And this new order will be updated in real time on the order book view.

4. Trade History: This is pretty straight forward. It's basically a view just like the order book but on the right side of the price chart. It shows a list of all the fulfilled orders on the platform.

If you have any prior knowledge on stock trading or other financial strategies and pattern analysis, GDAX is the best tool you can use for identifying opportunities. Also, if you want to keep a track of certain cryptocurrencies and where they're trading at, I would recommend a mobile app called Blockfolio (www.blockfolio.com). You can view your whole portfolio at once on your mobile which is very convenient. They also offer some management tools and price alerts.

Investing in Altcoins

Altcoins, as you must already be aware of by now, are cryptocurrencies that are not Bitcoin. As it happens, Bitcoin currently holds the lion's share of the crypto market. For many day traders and beginners, it may be overpriced or overwhelming. For such folks, there are a variety of alternative options available in altcoins which cost only a fraction of Bitcoin.

If you're interested in investing into an altcoin, you must be cautious of the hype that goes around before it's ICO (Initial Coin Offering). Most altcoins don't live up to the hype in the long run. So here are some tips to keep in mind before you put your money into an altcoin.

Tip #1: Check BTC correlations

Many altcoins have buy/sell trends that are directly affected by Bitcoin's pricing. When Bitcoin's value rises, the general tendency is a plunge in altcoin's value and vice versa. So, you must have a knowledge of Bitcoin's recent performance as well as a prediction of its future before investing in an altcoin. If you want to minimize risk, you should find an altcoin that has a decently high Bitcoin reserve.

Tip #2: Check backing and user-base

This goes without saying. You need to research about the goals, team, features of the altcoin.

Many altcoins have a cult-like following that you can capitalize on. For example, Dogecoin, which is a cryptocurrency that is literally inspired from an internet meme of a dog, has gone from a market cap of $60 million in 2014 to $580 million in 2017. And apparently a Dogecoin made out of gold is being sent to the moon in 2019!

Tip #3: Diversify your portfolio

Diversification is one of the best ways to minimize risk as an altcoin-investor. You should hold a reserve amount in stable altcoins that have a relatively decent future in terms of support and growth. Risks with ICOs and day trading should be done with a minor portion of your money that doesn't affect the rest of your portfolio.

Here are some more questions you should ask yourself before investing into a selected altcoin.

- What core problem is the altcoin attempting to solve and how big is it?
- Are there any alternatives available that address the same problem but more effectively?
- Is there a solid marketing strategy for the altcoin to reach target users?
- **Social Proof**: Are any other investors or VCs interested in funding the cryptocurrency? What are the tech moguls saying about it?
- **Buzz**: Is there sufficient media coverage (positive)? How many hardcore followers does the altcoin have on social-media?

- Is there a possible pump and dump scenario in the altcoin's future? How dependent is it on the founders and core investors?
- What has been the altcoin's performance in the past?

Chatper-7: Future of Cryptocurrencies

Cryptocurrencies have a smaller user base compared to regular fiat currencies. But as technology improves and more infrastructure and awareness are created around the world, the impact of cryptocurrencies is inevitable and immense.

Venture Capitalists have invested more than 1 Billion dollars into the blockchain technology itself. This is an indication of the scope of development that is bound to occur in this field in the upcoming future. Below is a graph showing the increase in number of user-wallets of bitcoin's blockchain.

Blockchain Wallet Users
15,271,519

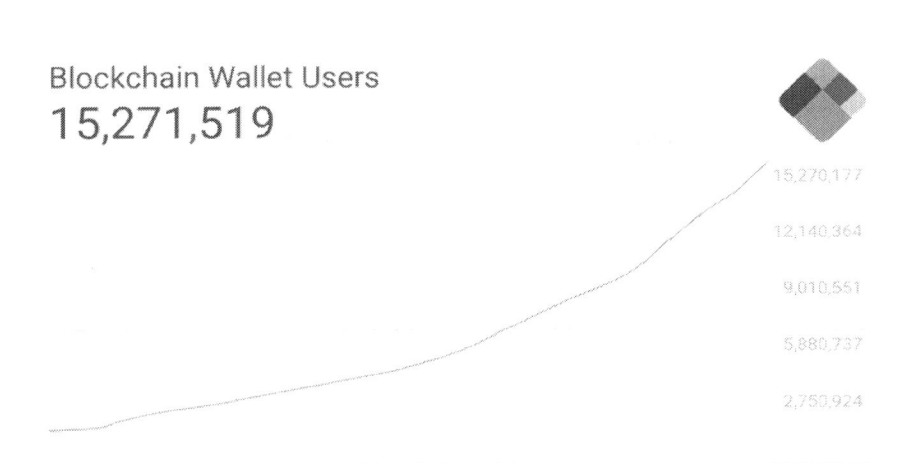

15,270,177
12,140,364
9,010,551
5,880,737
2,750,924

2013-07-29   blockchain.info/charts   2017-07-13

There are more and more cryptocurrencies coming into the market every month. As time goes on, we will see a range of cryptocurrencies offering different services for users. Bitcoin being the first one out there, will have an initial head start in terms of user adoption. But with all the latest innovations and the attention being paid to this space, it is difficult to predict whether bitcoin will be overtaken by other cryptocurrencies or not. As shown below in the graph from CoinMarketCap,

Bitcoin currently occupies more than 51% of the total market share of cryptocurrencies.

The advent of these new cryptocurrencies will be paralleled by an emergence of new crypto-exchanges. So, it will get easier for merchants and buyers to transfer the money and convert between two cryptocurrencies.

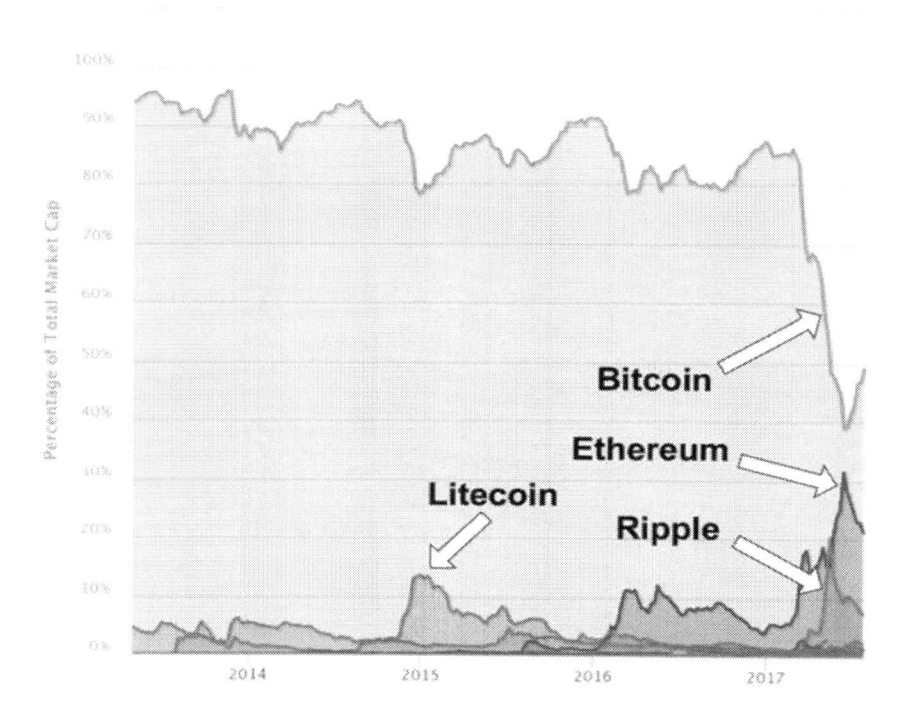

Having said this, I think we need to look at both sides of the coin (again, no pun intended). One of the biggest problems facing the mass adoption of cryptocurrencies is their lack of scalability. Bitcoin, for example, has had a huge growth in the number of transactions being carried out. The graph below, sourced from Wikipedia, shows how the number of transactions has been growing every year. But here's the catch – The block-size in Bitcoin is limited to 1MB. So, any blocks bigger than this are rejected by the network. This has resulted in limiting the number of transactions per second that can be processed by the network to three. To counter this limit, bitcoin miners have opted to upgrade the software so that the block-size can be increased to 2MB. This will increase the transaction fee but reduce the congestion in the network.

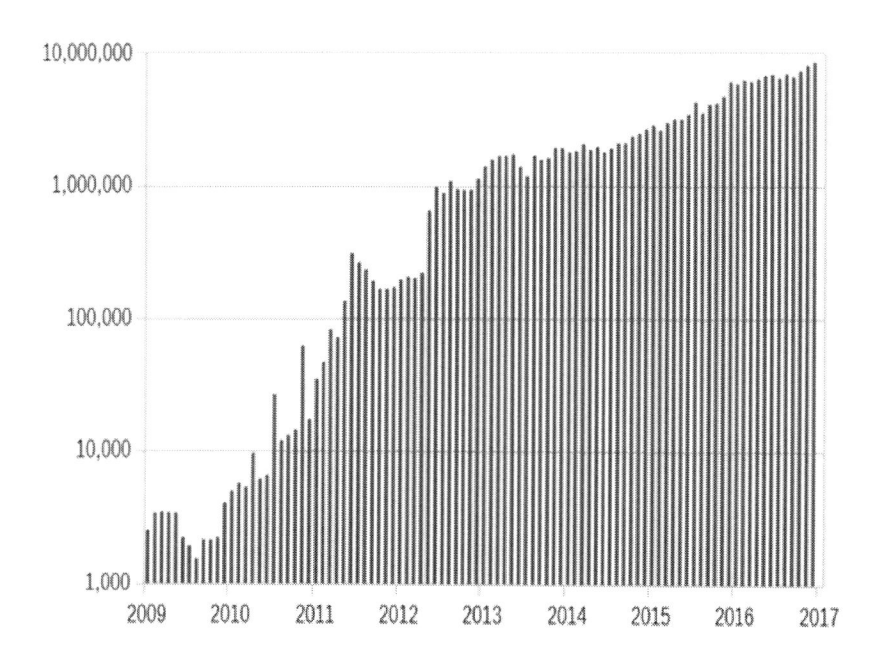

Bitcoin also had a bear market crash in the past (July 2017) where it's value dropped by around 20% in a 7-day period. Although the value is back up again, it would be foolish to believe that the cryptocurrency market is not volatile. It is highly advised for anyone interested in investing into cryptocurrencies or anyone that has already done so, to follow the latest updates and stay informed.

The Changing Landscape of Global Finance

With features like lower latency of transactions and reduced transaction fee, cryptocurrencies (especially bitcoin) have a potential for disrupting the e-commerce industry as well. The current online payment methods that users have to rely on for purchasing stuff online have lousy user experience, charge more per transaction and take longer to process payments. This gives cryptocurrencies like Bitcoin an opportunity to replace the traditional methods and create positive impact.

Due to its decentralized nature, Bitcoin has been facing restricted compliance from the banks and financial organizations. Its value dropped quite a bit when China banned Bitcoin from being used within its borders. But with blockchain, it's a different story. Financial institutions are showing positive response to the possibility of embracing the public ledger system. The reason for this seems to be the increase in operational efficiency created by using the blockchain technology.

Cryptocurrencies also seem to be advantageous for third-world countries that have under-developed financial infrastructure. These countries can bypass the need for spending a lot of tax money into public banks, mints and other regulatory financial organizations by directly adopting global cryptocurrencies like Bitcoin. What we're looking at is actually the possibility of unifying the world's currencies.

There are a lot of experiments being conducted in this space and despite all the hype, cryptocurrencies are still in their early stages. So, there is absolutely no need for you to feel like you're missing out on the party. Cryptocurrencies and blockchain have the potential to change not only the payment industry but also the way business is done. With widespread decentralized distributed digital currencies, there would be no need for separate national fiat currencies. All countries around the world could fall back on a single platform of value exchange. If it happens, this will be a landmark achievement in the history of human progress. Although we have a long way to go before achieving that stage, it is quite obvious that the future of cryptocurrencies is very bright indeed.

A few final words

Congratulations! You've made it to the end. Hopefully, it's been a fun and educational experience. I've certainly had a blast preparing this book for you. And once again, I want to express my deepest gratitude to you for having given me your time and attention. I hope you found great value worth your investment. If you did, I want you to just do this ONE thing for me.

Please leave an honest review on Amazon at the link below. That is the only way for me to get your feedback and improve my craft. Thanks a ton!

www.bookstuff.in/devan-hansel-cryptotrading-review

More books from the author

Bitcoin: The Digital Gold

Blockchain: The Technology Revolution behind Bitcoin and Cryptocurrency

Cryptocurrency: The Essential Guide to understanding Bitcoin, Blockchain & More

Printed in Great Britain
by Amazon